THANK GOD FOR THESE GUYS

CHICAGO'S FIREMEN ON THE JOB

PHOTOGRAPHY BY **ALAN JACOBS**

M.T. Publishing Company, Inc.
P.O. Box 6802
Evansville, Indiana 47719-6802
www.mtpublishing.com

Graphic Designer:
Alena L. Kiefer

Technical Photo Consultant:
Krysia Jacobs

Library of Congress Control Number:
2007921567

ISBN: 1-932439-52-8

First Printing: July 2007
Second Printing: November 2007

Printed in the
United States of America

01-08

THANK GOD FOR THESE GUYS
Introduction by Alan Jacobs

Waiting to be discharged, I am putting on my pants at Northwestern Memorial Hospital in Chicago on September 11, 2001 at 8:02 A.M. I pull on my right leg while glancing distractedly at the TV on the wall and see the North Tower of the World Trade Center in roiling flames. I think: "there must be some sort of electrical fire," never dreaming of the reality. I stop and watch the thick billowing black smoke while putting my foot in the other leg and pulling up my pants. Looking up at the screen again I see, or rather witness, the South Tower explode into flames at 8:03 A.M. I have no idea UA 175 had crashed into it. I think the heat from the North Tower must be so intense it ignited the other one. I ignore a black speck flitting across the screen from the left, simply denying the reality. Seeing the replay a few moments later, I realize what that spec was and what has happened. In the time it takes me to pull on my pants 343 firefighters, 75 policemen, men and women, and 2334 civilians gone; little kids empty hands, husbands, wives and lovers beds, driver's seats, bunks in the firehouse, seats on the rig, clothes in the closet, all empty, kaput, their occupants gone forever.

Several minutes later sitting in a waiting room, I see the first tower collapse. It just comes down... woosh... like a controlled implosive demolition of an old building. Terrible and fascinating to watch, right? I can't look at one now without thinking of that day.

Afterward I percolate for about 10 months, realizing I know nothing about what firemen actually do, day-in, day-out. Four things keep nagging: deep rage for the murderers; empathy for the murdered civilians and their families, sadness for the firemen and policemen who gave their lives, and for their families; and an enormous curiosity about what it is firemen actually do. I realize I know next to nothing about what their work is like and very little, if anything, about the bond between them. The best I can do at the time to show my sadness and respect is to bring pizzas to the firehouse near my home for several months, not leaving my name. Each time I get back in my car in tears. In light of 9/11's devastating reality my actions, taking little effort or time, are next to meaningless. I feel somehow guilty. I owe them something, something personal, something owed for a long time. I am searching for some way to stop standing on the curb watching a fire, or leering at bodies lying in the street after a bad auto accident. At some point I stop being a bystander and decide to give something back to those who lost their lives that day attempting to save others, and really to those people all over the country sitting in their firehouses who could have been in those buildings on that day. Thus the idea for this work was hatched.

This is not only a site for firemen, men and women. It is a about them, about what they do all the time. It is about their work and their tradition and dedication. It is for all of us who don't know. It is for all the people to whom I have shown photos who never realized how curious they actually were. We can honor the memories of those fallen in the line of duty with ceremonies and dedications; not only at the Trade Center but all over the country, on average 116 a year. But it isn't enough to just feel badly for them. What they would like but hold out little hope for is to be understood and not taken for granted, the very thing all of us need. So I set out to discover as much as I can, never realizing at the time the complexity of the job or how captivating it would become, the tight-knit community I was allowed to enter, the tragedy, and the humor, the honesty and openness, the help I would receive and... the friendships. What a bunch they are.

3

Forewords

I first met Jake when I was the Director of Training for the Chicago Fire Department. Jake came into my office, introduced himself, and explained a little about his project. I told Jake his request to "ride along" and take pictures would have to be approved by the Fire Commissioner. I gave him contact numbers to pursue his ideas

I got to know Jake personally when he was doing his ride-along with Captain Mike Cahill of tower ladder 5. At that time, I was assigned to Battalion 4 and Tower Ladder 5 was under my command. Mike Cahill was not only the Captain of Tower Ladder 5, but also the union business agent for Chicago Firefighters Union Local 2 first district. Mike was a well-respected fire officer and a union official so when Mike said Jake was "OK", his word held meaning.

Jake explained his interest in what we do and how it came about. He was recovering from a heart procedure when the 9/11 tragedies occurred. His curiosity peaked as to just what kind of individual would want to be a firefighter or paramedic, and so began the book.

I think his endeavor to show the faces of those who protect life and property, is a great way to make the public aware of the work we firefighters perform, and the character we truly have. Jake has seen first-hand the brotherhood and sisterhood of our life in the fire service. He has spent countless days and sleepless nights out on the streets of Chicago witnessing our dedicated and courageous firefighters and paramedics. He has observed the physical and emotional side of this profession.

Our Fire Commissioners, past and present, have asked Jake to capture photos of training evolutions and special events.

This book is for civilians, firefighters and paramedics. It tells a pictorial story of our lives and how our profession is our life.

Jake's work has taken him as close to the "real thing" as anyone can get. Some of the pictures depict the tragic incidents we encounter daily yet are considered too graphic for public viewing.

We are grateful to Jake and his wife Krysia for their hard work in assembling this book.

Thank you,

Robert S. Hoff
Assistant Deputy Fire Commissioner
Bureau of Operations

Firefighters and Paramedics are a unique breed. We encounter people on perhaps the worst day of their life.

We look at each other and find strength, not weakness…….. Faith, not fear…….. Unity,…… not division.

We have all heard the phrase, "tomorrow is promised to no-one". Yet, each and every day we wake up and step into a uniform that represents hope, courage, and compassion. We do not know what the next hour holds, but we continue to respond. For that I say "Thank You" to the men and women who choose to serve.

They are greatest people I have ever had the privilege and pleasure of knowing.

We grieve for the lives lost and the loved ones left behind. We honor their memory, and the memory of all those whose names are forever memorialized and etched in our memory.

Raymond Orozco
Fire Commissioner, City of Chicago

Acknowledgements

When I read a book, I usually skip this part, finding it tedious and boring. And I ask you, please, tolerate this one because, quite honestly it is mine, and more importantly these guys deserve it.

First is Chicago Fire Department Assistant Deputy Commissioner-Operations Robert Hoff. Without him? Well no book. Hey Bobby, thanks for the stories, introductions, invitations, brain-power, openness and sharing, advice, and most of all pal, the bunker jacket and door wedge, real and figurative. Next its Ray, as he insists for all, CFD Commissioner Ray Orozco who walked up to me on the street about four years ago, when he was a Deputy District Chief, put an arm over my shoulder and took me into my first overhaul. Four years later I was photographing a huge demonstration the first day Ray was on as CFD Commissioner. There was an isle cleared in the middle of a crowd of well over 200,000. He spots me entering it, comes up and puts his arm around me; like he didn't have anything else on his mind. Same Ray, always the same: kind, brilliant, passionately dedicated to saving lives, and funny as hell. And Patrick me-boy, Lt. Pat Lynch, for always being exactly who you are and for teaching me so much: FYYFF... Thanks also to two men who have shared many stories, professional and personal: 5th District Chief Cortez Holland, and 1st District Chief, Jose Santiago.

Large thanks are due to former CFD Commissioner Cortez Trotter, now the Chief Emergency Officer, Chicago, who asked me to shoot several CFD functions and believed in the project enough to give me a CFD I.D. as a photographer.

Also a big thanks to former Commissioner James Joyce for taking the time to really look at my work, share his stories and photos, and give me a riding pass. Thanks are due to Head of Special Operations, District Chief Mike Fox who allowed me to ride with his boys on Squads 1, 2, and 5, and to proudly wear the Special Ops patch on my shoulder.

Thanks also to Battalion Chief Rich Edgeworth for dragging me along in the buggy. Edge, "It's a beautiful thing," buddy. Thanks to 14th Battalion Chiefs Jerry McKee and Mark Ward, and the boys at Engine 38, Truck 48: Lt. Mike Samandra, Capt. Juan Reyes, and Angel, Joe, Jimmy, Marty, "Stash" Mo, John, also to Lt. Jack McKee, Capt. Mike Cahill and all the boys at Tower 5, for squeezing me in that enormous rig and helping quite a lot: Jamie Gonzalez, Larry Garza, Raudel (Rudy) Cassanova, Arlency Pitts, Jesse Sanchez, Vinicio Espinosa, FF-Paramediac Kevin Smith, John Knightly, Arty Jansky, Pitt, Dave Bautista, and Capt. Kevin Bernaciak. Thanks to Chief Bobby McKee for letting me ride in the "buggy" and for sharing a bit of his enormous knowledge and sense of humor; thanks "boss". Thanks to Lt. Cliff Gartner, Capt. John Strong and the boys at Squad 5: Relief Lt. Will Trezek, Corey Hojek, Steve Groszek, Tommy Meziere, Brian McArdle, Bobby Smith, Tommy Garswick, Shun Haynes, Jamar Sullivan, Pat Noonan, Joe Atkins, Glen Keyes, Brain Velez, Brian Herrli, and Billy... Billy O'Boyle for all the things you guys have done for me, more than I can ever repay. Thanks Engine 116 Capt. Sean Burke, and all the boys and girl, on that vehicle.

Then there is Squad 1... all three shifts: Capt. John Collins for explaining so much... Capt. Jimmy Altman for all the understanding, Lt. Bill Duffy, and relief Capt. Kevin Krasneck for sharing considerable knowledge, and "the guys": Brandon Dyer ("The Pearl"), Manny Soto, Dave Schick, Danny Truesdale, Tony Budvaitis, Tom Dati, Kelly Burns, John Scheurich ("Scooter"), Pat McCauley, Jim Stepien ("Step"), Dave Leon, Joe Kubik, Jay Lopez, Bob McCrea, Pat Brown, Raul Ochoa, John Haring, Dave Gates, Mike Chesack, Mike Murphy, and the inimitable John Schienpflug.

Thanks to Steve Curley and his brother Lt. Pat Curley for suggesting and receiving me on the West Side of Chicago.

And thanks to Deputy District Chief (1st) Eddie Enright; thanks for all those nights up there in the 1st's office "Double E," sharing stuff about your time on the squad, the guys, the pride in the job, and the Nam.

And a special thanks to 1st District Chief John Brooks, 1st Deputy Commissioner, for putting the bug in my ear about doing work for the CFD. Thanks also to the CFD photo unit, and to its erstwhile leader Randy Clay.

And...and... and... Krysia Hnatowicz Jacobs, compuwiz, partner, collaborator, technical brain behind it all, sensitive, brilliant, caring beyond human imagination... my wife of 27 years. Thanks pal.

"Jake," Chicago, 2007

Author's Note

The term fireman is not intended to exclude the women in the fire service. Instead it is meant to include them in a time honored tradition, using the term generically as alderman, actor, soldier, director are used. Moreover, every female "firefighter" (I asked eight) said they preferred "fireman."

A picture is worth how many words? Fireman Billy Milton of Truck 7 out of the fire building at 19th and May, 1/14/04.

(Opposite Page)

Lt. Kevin Bernaciak (L) and fireman Dave Bautista, both of Engine 23, were ready to climb in the window of this 1½ story brick home on 15th St. when it flashed, forcing them to temporarily back part of the way down the ladder. Bautista was able to put water on the fire with a 1¾" hose, seen trailing behind him on the right. The fireman in the foreground is Lt. Mike Smandra of Engine 38.

(Above)

As you can see, Bernaciak and Bautista progressed, the hose now in the window. When I first was allowed to go in when it was safe enough, I couldn't understand how these guys took the smoke. It's a matter of willpower and conditioning. Today's firemen refer to the men before the advent of O₂ equipment as "leather lungs".

(Opposite Page)

(Above)

Battalion Chief Jerry McKee's birthday party at the firehouse. A former SS1 (Snorkel Squad) member, his command philosophy empowers his men allowing them do their work, instilling confidence in themselves, and him. This is really essential in stressful, life threatening situations. Love is not too strong a word for what his men feel for this smart, knowledgeable, modest, and very funny man.

(Above)

Throwing-up the Ladders
An important aspect of tactics is ventilation, that is, cooling off the hot gasses inside and clearing the smoke. Windows are broken, doors and the roof are opened. One way to get to the roof is by ladder. Two sizes are being used here. The men in the center are struggling with what is known as a "banger", the largest "ground ladder" used by the CFD. It is, according to one knowledgeable Chief, F____n' heavy. Fully extended it can be raised 50'. The fireman in blue with his leg raised is Relief Chief Pat Knightly. Other ground ladders used by the CFD range between 28' and 45'. This was a still-and-box alarm, 2/7/03 on W. 15th St., just east of Mt. Sinai Hospital. A middle-aged woman failed to make it out.

(Opposite Page)

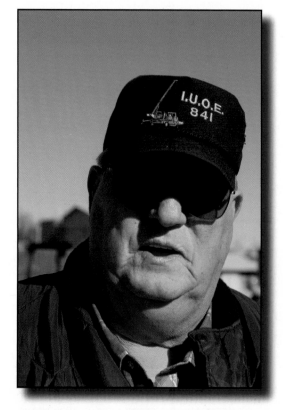

"Wes"

This homeless man irately refused help. Paramedic-fireman Gary Michalek attempted without success to convince him to at least go to the hospital.

(Above)

3-11 alarm on N. Wells St. 1/07/05. The basket directly overhead is a Tower Ladder, whose straight arm can be extended 100ft (approximately 10 stories). The basket behind it is the snorkel of Squad 1A. Fully extended, it reaches 55 ft., but it is never extended to its full height for safety reasons, avoiding the possibility of tipping the truck on its side. Unlike the tower ladder, its arm is hinged allowing greater maneuverability in tight spaces. The men in the white hats are battalion Chiefs and the man in the white coat and hat is a Deputy District Chief. Behind him with his hand on the parking-sign, is 1st District Relief Chief, Keith Witt. This was a dangerous fire very difficult to defeat. In this instance confining the fire to one building was considered a victory.

(Opposite Page)

This is the rear of a vacant single-family home. Fire doubles every minute and this fire was aided by a strong wind. It pushed it right through the building from the rear to the front. Eventually the brick wall in the rear collapsed. I recall taking my camera down and just looking, taken as I was by the fire's incredible power and beauty.

(Above & Opposite Page)

Smitty

Squad 5 fireman Bobby Smith hit by a fragment of lath he chopped off the wall with an ax during overhaul on 7/10/04. He hardly remembered it when given the photo.

(Above & Opposite Page)

L abor Day '02
Engine 38's guys opened the hydrant with a bell fog nozzle for these kids on a sweltering day in the Latino neighborhood known as Pilsen.

(Above)

I t's not all about towering flames, flashing hoses and axes, and people working swiftly against death. Fireman /EMT Tracy Scott of E. 103 at a summertime street fair.

(Opposite Page)

Engine 116 Fireman Todd Taylor during overhaul at unidentified fire.

(Above)

Squad 1 Fireman Raul Ochoa, a recipient of the Carter Harrison Award for Valor, rescuing a man from an upper floor by securing him with body and arms, at a 4-11 alarm fire at 2121 W. Washington St. on 2/11/03. Every time I see a guy climbing a ladder, in this case on a +9° night with 20 mph winds (-9° wind chill), I think to myself, "what kind of men are these? What are they made of? What drives them to do such things?"

(Opposite Page)

Corey Hojek of CFD Squad 5, after a fire on Chicago's South Side Dec, 18, 2003. Hojek is also a shift chief in Evergreen Park, a Chicago suburb. His father was chief of that entire department. Three of Hojek's brothers are also firemen in various Chicago suburbs.

(Above)

Deputy Commissioner-Operations Gene Ryan (L) and then 4th District Chief John Brooks, two knowledgeable firemen and leaders, directing fire-ground tactics at a 4-11 alarm fire on 26th and St. Louis, 1/21/06.

(Top)

Larry Anoman of Tower Ladder
34 after a fire on 11/26/05.

Tower Ladder 5's aerial and basket in the smoke at 4-11 alarm fire, 11/12/02. In addition to putting water on the fire from above, this provides a more global view, the information radioed to Fire Command on the ground. The Tower is linked to a $2\frac{1}{2}''$ hose that gets water from the Tower's pumps traveling up the main ladder to a nozzle on the front of the basket. The water for the pumps is received from an engine attached to a hydrant.

(Above)

Doin' His Thing.
CFD Fireman John Knightly, now a Lieutenant, in the box of TL-5 at a 4-11 alarm fire in 11/12/02. This piece of equipment is used for several things, including getting above a fire and pouring water down on it, rescuing people from windows, roofs, perches etc., and observation of special events like a rally or march of some kind.

Modern Iwo Jima.
From right: Tony Budvaitis and Danny Truesdale, and (...) at an extra-alarm fire on W. 22nd St., 11/12/02. Fires such as this provide an unlimited amount of subjects. Everywhere one turns, there are men and equipment engaged in "the work". The men in this photo remind of flag raisings at WW II battle for Iwo Jima on Mt. Suribachi, and at WTC Ground Zero, 9/11/01.

(Above)

Angel over the City.
Tower Ladder 5 at the same fire on Cermak Rd. The spray emanating from the underside is not for extinguishing but rather to cool the men in the basket.

(Opposite Page)

The Basket of Tower Ladder 5 in action at an extra-alarm fire on 19th and May, 1/15/05. Tower Ladder 5 firemen are tearing away part of the roof, searching for more fire or smoldering embers. They eventually used the water cannon at the front of the basket.

(Above)

View from the top. Tower Ladder 5 fully extended.

(Opposite Top)

"The Basket" of Tower Ladder 5.

(Opposite Bottom)

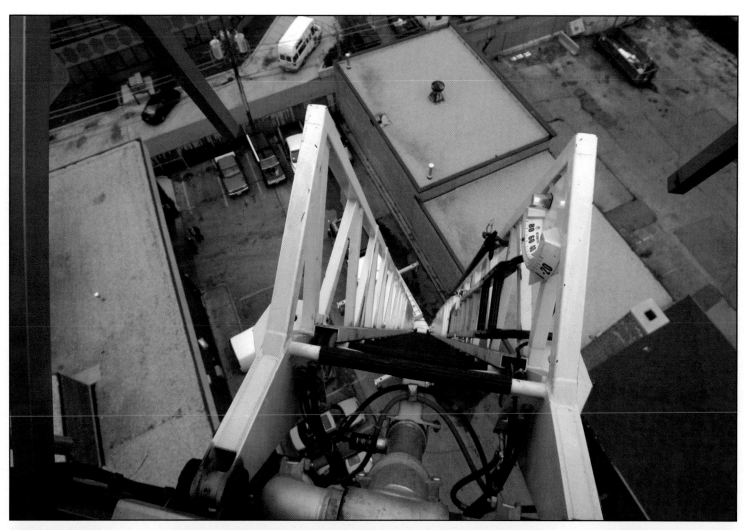

Cleaning the hoses.
Most pitch-in and do their share. Pulling your own weight even on these kinds of tasks is a reflection of how you will be in a tight situation in a fire. This is the wisdom of firehouse culture. What each man wants to know is the bottom line, whether the other man will hold his own and, if need be, pull him out if he goes down, the same thing combat troops need to know.

(Above & Opposite Page)

H omeless in The Park. In Chicago, an engine, a truck and an ambulance, 10 firemen and two paramedics or EMTs might very well be called to such an incident. The amount of time, physical resources, manpower, and dollars, spent for this one down-and-out soul is something few of us civilians understand.

(Above)

T his young father in his middle 20's choked to death on some food. Here, Engine 83 Fireman/Paramedics Mitch Ludwig and Anne Cacioppo (blue gloves) intubating their patient that is, inserting a tube in the trachea (windpipe), for ventilation.

(Opposite Page)

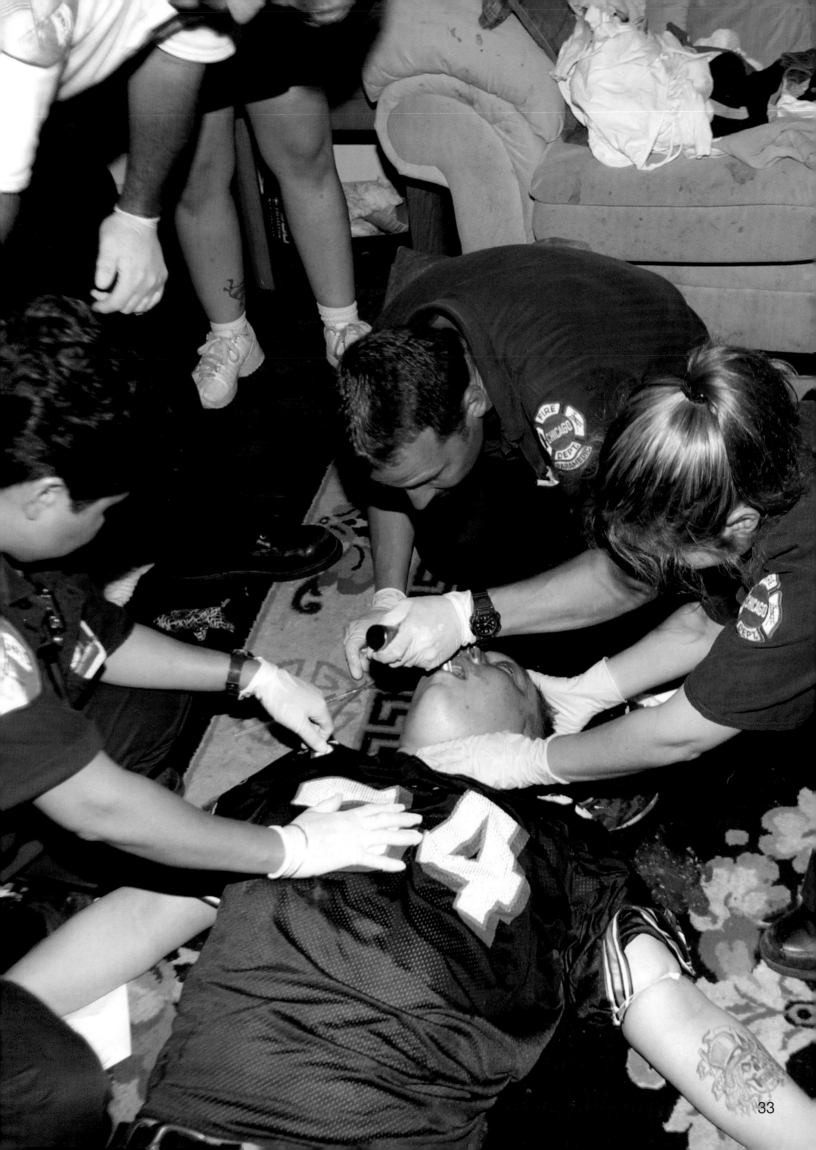

A Rainy Night
Squad 5 fireman Joe Atkins, operating the Hurst Corp.'s Jaws of Life™ cutter at an auto crash 7/10/04. A portable gas operated power motor sending pressurized hydraulic fluid to the cutter powers this device. The supervising officer with the light is Squad 5 Capt. Eric Strong. The damage was so extensive, it took over a half hour to free the two young women pinned inside.

(Above)

"Car into Building" is what these types of accidents are called over the speakers, this one at 7:30 AM on a Sunday morning. The young man driving was racing someone down Garfield Blvd. at approximately 70 mph.

(Opposite Page)

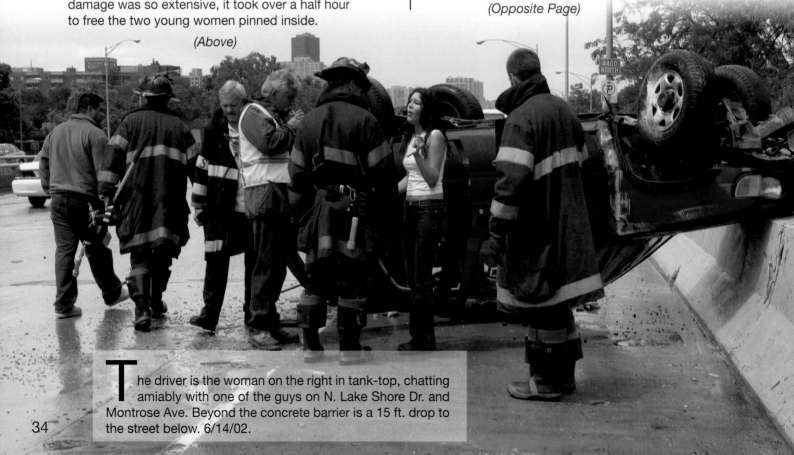

The driver is the woman on the right in tank-top, chatting amiably with one of the guys on N. Lake Shore Dr. and Montrose Ave. Beyond the concrete barrier is a 15 ft. drop to the street below. 6/14/02.

Looking for the Beast
They are searching for more fire: embers between pieces of flammable material, under roofing material. It was 5°.

(Above)

Ominous Sign
The white smoke flowing from the far side of this single family home is actually steam created by water hitting flames on the first floor. The black smoke emanating from the attic window is a sign of hot gasses that almost certainly will burst into flame.

(Opposite Page)

Early stages of a 2-11 alarm fire at 6620 S. Aberdeen in Chicago, 12/17/05. This shows the fire communicating to the structure in sector 2. Sectors are designated from front and clockwise around the building 1, 2, 3, 4. Here, the fire in the primary building in sectors 1 and 2 is leaping across to the next building's sector 4. Any buildings next to the fire building are designated "exposure buildings."

Unidentified firemen at Still and Box 9/5/03.

(Opposite Page)

Lt. Jimmy Altman Squad 1, now a Capt. His two brothers, also captains, are on the CFD and his father was a former Commissioner.

(Above)

T ruck 48 Fireman Joe Tito taking a water break at a Still and Box, 12/06/02, on W. St. Louis Ave.

(Above)

L ife-Long Buddies.
Lt. Pat Lynch (left), and then 6th District Deputy Chief (DDC) Ray Orozco clowning after a fire on Chicago's Southside, 2/14/04. Both are deeply involved in training to save firemen's lives. As of this writing Pat Lynch is now on Squad 2, and Ray Orozco rose from this job to District Chief, then to Assistant Deputy Commissioner-Operations, and then to CFD Commissioner. "Ray" has been busting Pat's b---- for years about the size of his nose. Firemen's interaction with one another reminds one of sports locker rooms, with their good-natured bantering and practical jokes. The difference is losing has much more serious consequences.

(Left)

J ohn O'Malley "pulling hose" full of rushing water that is extremely heavy and relatively inflexible. The people at the nozzle, or "pipe" end cannot advance it by themselves. Putting water on a fire is important not only to save property but also to save firemen's lives. Advancing on flames entails risks, not the least of which is a flash or rollover that can envelope, severely burn, and sometimes kill firemen.

(Opposite Page Top)

Memorial Day 2006, Holy Family Church. Commissioner Orozco: "The woman I'm hugging is the widow of Lt. Thomas O'Boyle who died in the line of duty on September 26, 1995. Tom was a very close friend of mine, kind of a second father to me; we worked together on the South Side & West Side. Pat is a member of the Gold Badge Society. Tom also worked with my father on E-84 and I have known him and his family since I was 12 years old." [The Gold Badge Society is comprised of families of firefighters and paramedics who made the ultimate sacrifice, losing their lives in the line of duty.]

(Left)

A class visit to the firehouse.

(Above)

I Wanna be a Fireman When I Get Big. Kindergarten children coming to the firehouse on 10/9/02, less than a year after the World Trade Center catastrophe, with fire helmets made out of construction paper. I asked one of them if he knew what happened at the Trade Center and he said, "when the building came down and all the people were 'kilded.'"

(Opposite Page)

A Family Tradition
Lt. Jack McKee of Tower 5 (L), and Battalion 14 Chief Jerry McKee. Their father was a Chicago firefighter, as is Jack's son Dan. His daughter Tracey graduated in the Spring of 2004 from the Oak Lawn Fire Academy At this fire, Jack's men manned Tower 5 and played a major role in extinguishing the fire from above. Jerry's Battalion 14 men were also very important, entering the building from both front and rear. The concern on their faces occurred after Chief McKee was forced by conditions to call a "Mayday". There were some anxious moments. Luckily everyone got out.

T he McKees

It took me a year to get all these people together, with the different shifts and all. From left: Chief Bobby McKee, Lt. Jack McKee, fireman Tracy McKee, Chief Jerry McKee, and fireman Danny McKee. Bobby, Jack and Jerry are brothers. Tracy and Danny are Jack's kids. Both Jerry and Bobby have sons now in the fire service. In all, nine McKees have been firemen, including their father.

CFD Squad 5 Fireman Brian McArdle calling for a hose line at a Still-and-Box on the far South Side of Chicago, 2/7/04.

(Above)

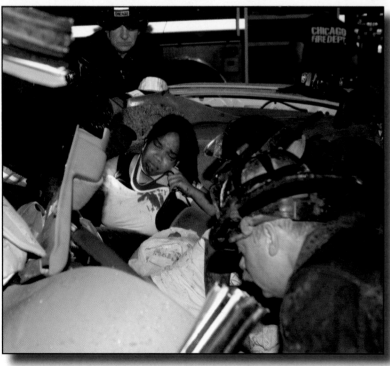

Cutting all four posts and removing the roof freed these two young women from their car. It's something most of us don't give a thought to, the amount of time, energy, manpower and equipment deployed in these accidents.

2 Jo-Jo
Joe Atkins and Squad 5 teammates opening, or venting, a porch roof in the dead of winter.

(Opposite Page)

D an Lanham of Engine 109 one cold
Saturday morning, in January '06.
-6° (-15° windchill).

"Hollywood" Steve Curley of Truck 47.

Cold day, -6° (15° windchill). Fireman Frank Sanders of Engine 107.
(Above)

Box Alarm on the far South Side of Chicago, 2/7/04.
(Opposite Page)

CFD Squad 1 fireman Jay Lopez in the late stages of the overhaul process. Even though thousands of gallons of water are used, fire might still be smoldering in ceilings, walls, doors and window frames; in short, in any concealed space or furnishings. Given enough time, these apparently subdued areas can erupt or, as it is known in the fire service, "rekindle". That term indicates a fire that ignites anew after everyone has left, requiring a return, often hours later.

(Above)

Firemen often refer to their enemy as "the beast." It is the primary destructive force in nature. It is that, to be sure, and it is also the source of all life as we receive this proud gift of energy from the sun. Here at a 2-11 on N. Wells St. in Jan. of '05, a relief of the devil seems to be laughing down at those trying to defeat his latest destructive eruption.

(Opposite Page)

On the street, in the cold, watching their homes destroyed by fire.

(Above & Right)

The woman was watching her apartment burn, had escaped with her baby just in time. Think of it: you are in the street. It's cold and you and your baby didn't have time for warm clothing, not even shoes. All your possessions are gone. You don't have much money. You don't know where to go or what to do next. All you know is you are really scared. Luckily, there is a Human Services Dept.

(Opposite Page)

Squad 1.
(L-R) Jay Lopez, David Gates, 1st District Relief Capt. Kevin Krasneck, Mike Chesack, Mike Murphy and Pat Brown.
(Above)

The Boys From Squad 5 Second Shift
(L-R) Brian McCardle, 5th District Reliefe Lieutenant Will Trezek, Corey Hojek, Bobby Smith, Tommy Meziere.
This photo was taken when I first started riding with these guys. The Squads in Chicago, 1, 2, and 5 are what might be called the Navy Seals of the Fire Department. For one, they respond to fires over a much wider area than the regular trucks and engines. Their other responsibilities are HAZMAT, water rescue, scuba diving, repelling, high angle rescue, vehicle accidents when people are pinned-in, and confined space rescue. This is a wide and varied range of tasks requiring a great deal of skill and training. There have been days when I was along on five fires. Depending of course on the sizes of the fires, two fires in one day are considered very good by the regular trucks and engines. Not to make light of the physical condition of other firemen, the squads are in great shape. It must also be said that there are many fireman on other trucks and engines who are their equals.

(Opposite Page Top)

Squad 5
(L-R) Brian McArdle, 5th District Relief Lt. Will Trezek, Brian Herrli, Shun Haynes, Steve Groszek, after a still alarm on the South Side of Chicago, 12/21/03.

(Opposite Page Bottom)

4-11

alarm fire at 2121 W. Washington St. on 2/11/03, 20 mph wind, 9° (-10° wind chill), with water flying everywhere. Everyone and everything was covered in a layer of ice. It was so thick that when I tried to climb into Tower Ladder 5's cab I couldn't bend my leg to reach the first step. I had to hit my legs with my fist several times to break up the ice. Once inside, the ice melted just enough for me to realize I was wet. When I climbed down and the wind hit me, I got cold. But as soon as my jacket and pants froze again, I realized what the Eskimos have known for centuries: ice is a great windbreaker.

(Above & Opposite Page)

Squad 1, 3rd shift, about ready to go. Fireman Manny Soto appears to be pointing and describing something to Lt. Bill Duffy. This group walked up to the fire seen in photo on right, put a hose on it, and then went in. It seems so simple, sitting here writing this...

(Above)

The Mouth of the Beast
This was a 2-11 alarm fire on W. Monroe St. 2/5/03 in a 2 1/2 story residential building. Not too long after this photo was taken, they walked right up to this roiling inferno with the hoses, pushed it back, and went in.

(Opposite Page)

Father Tom Mulcrone's helmet and mask at a fire on a hot summer day.

(Above)

Father Tom

Mulcrone,CFD Catholic Chaplain, at a 4-11 alarm on 26th and St. Louis, 1/21/06.

(Opposite Page)

Less than a year old in July 2002, this fire truck is used to extend its ladder as high as 100 ft. or approximately five stories in rescue attempts. It also goes on medical emergency calls. Up front ride a driver and an officer, either Captain, or Lieutenant. Three firemen, one of which can be a paramedic, and one or two who could be EMT's (Emergency Medical Technicians), ride in the cab behind the driver's compartment. Truck Company 22 has been in existence since 1893.

(Above)

The highest part of the scene is the lowest in the photo. This, at a still-and-box alarm on S. Lituanica Ave. 1/14/03. Notice the fireman stepping onto the roof.

(Opposite Page)

Hose Tower in one of the new firehouses at 5th District headquarters, 59th and State on Chicago's South Side. This is used to hang wet hose for drying. Unlike the old towers, this type utilizes an exhaust system.

(Above)

Firehouse Sculpture: E-109.
These antique spiral staircases were built to prevent horses that pulled the engines from climbing the stairs to reach the hayloft.

(Opposite Page)

E xtra alarm on S. California, 3/3/04.
When it snows, most of us civilians groan and moan about the inconveniences - wearing boots, having to shovel sidewalks, being splashed by passing cars, starting our cars, clogged side streets, personally shoveled parking space conflicts, traffic delays, etc. Imagine what it must be like for these guys: frozen hydrants, wet gloves, slippery sidewalks and streets, difficulty getting the vehicles in the proper positions, icy ladders, icy snow covered roofs, and when you get wet you can't just stop and dry and get warm. They have to deal with the discomfort, even suffering. In this respect, it's not unlike combat. They just deal with it.

(Above)

5° Below and Still Working.
The device attached to the hydrant is a type of "gate valve" that regulates flow, thus enabling two hoses to be running at once or to alternate between them. In this photo the connection is facing and closed.

(Opposite Page)

Louie-the-Lob

Louie was a famous character known by many in the fire department, even by several CFD commissioners. Firemen are like the sabra cactus found in the Middle East, tough and thorny on the exterior, sweet and soft within. Often, they adopt people who are down and out in some way; a lonely soul whom society has all but rejected, an intellectually or physically challenged person, a retired guy living in a nursing home that comes by for some real food and a cigarette. As tough and hard-nosed as they can be, many firemen possess an unusual degree of humanity.

In approximately 1930, when Louie Cerveny was 4, his brother was told to take him over to the firehouse less than a block away. Louie would spend all day there almost every day. Some kind of mutual bond was forged rapidly. He attend school until age 9, when he had a grand mal epileptic seizure in school and a frightened teacher locked him in a closet and forgot about him. Worried why Louie hadn't returned after school, his mother went there only to find him still locked in the closet. Louie was so traumatized by this event that his mom withdrew him from the school, all school... permanently. Sadly, he never returned and therefore never learned to read. After the school incident, he was raised by the guys, only returning home to sleep. He ate breakfast, lunch, and dinner with them on all three 24-hour shifts, and one of them would walk him home in the evening. At some point, medical authorities said he had to go to a home for juvenile delinquents. He was there for some months. After this he continued going to the firehouse.

When he was old enough, Louie went to work on a beer truck delivering 300 lb. wooden beer barrels. He didn't ride on the fire trucks very often, preferring to stay in the fire house. Every morning he would go shopping with whomever was cooking for the day and would buy the coffee and sugar.

He was a sensitive, charismatic, irascible, outrageous, and lovable cuss, full of profanity and an occasional whack on the arm. His eruptions were famous. Once he punched Battalion 14 Chief Jerry McKee in the arm so hard McKee's arm was black and blue for a month; a form of affection from Louie.

He met his pal, and eventual caretaker Bob Molinari around 1980. Once, when he met Bob's mother he erupted with f--- you, you c---, a typical way of testing if she was safe.

Because of his increasing physical problems, not the least of which was dementia, (now a Lt.) Molinari took Louie in to live with him. Louie chose this over the offer of his nephew. Bobby fed him and bought his clothes, and gave him baths - not without a good deal of resistance - once a week whether he needed it or not. And he took him to the firehouse every shift he worked. In the last year or so, Bob dealt with Louie's increasing incontinence, and dementia. He says, "Louie was like a father, a son, a best friend, and in the end, more like a kid." He wept openly when he described finding Louie in a chair outside the firehouse... gone. Louie had been with his firemen for 73 years. His wake in Chicago was jammed, including several former commissioners.

Louie the Lob with just a few of his guys.

(Above)

Louie's pal, brother, son, and father, Bob Molinari of Engine 72 at Still and Box on S. Cregier 4/28/06. His O^2 tank ran out and he was hanging out the window for some air, and calling for a ladder.

(Opposite Page)

Steve Groszek

I t's not double vision. The new (right) replacing the old.

T he biggest heart in the CFD. Capt. Mitch Crooker (center) at Illinois Fire Service Institute in 2003.

(Above)

A Still-and-Box on Chicago's South Side 11/19/04.

(Opposite Page)

74

"Bri"

One of Bobby Smith's team mates on Squad 5-2nd shift, Brian McArdle, at the fire on 47th and Princeton, 2/28/04. If I couldn't find Bobby, I'd look for Brian.

(Above)

"Smitty"

When doing this writing, I couldn't wait to get to this guy, Bobby Smith. He's on Squad 5. Bobby, funny, sometimes painfully honest, was the man I looked for when they would let me in for overhaul, the last stages of fighting the beast. He was/is almost always in the middle of it, working his butt off, snorting, sweating, nose running, caked with wet wall and ceiling debris. He took the time to explain many aspects of the job.

The archival photo below is that of Bobby Smith's great-great grandfather (left), also a Chicago fireman. *Archival photo courtesy of Bob Smith.*

(Opposite Page)

Pulling ceiling during overhaul: much of the work at a fire comes after it is apparently extinguished. Very often this procedure reveals hidden embers and flames.

(Above)

Better Safe Than Sorry.
Civilians sometimes complain how destructive firemen are. The ceiling at left looked just like the part of it on the right 30 seconds before. Often, fire only seems extinguished. On further poking in ceiling, walls, floors, doors and window frames, it is often discovered smoldering away in hidden places.

(Opposite Page)

Two young women got out just in time in the Pilsen neighborhood on the west side.

Watching their house burn on the south side in the Englewood neighborhood, part of Chicago's enormous "ghetto." Very often this means people have literally lost everything.

Smitty and "The Hurst Tool"
The Jaws of Life™ are hydraulic rescue tools, developed by Hurst in 1972; spreaders, cutters, rams, airbags to lift heavy objects, and other devices. Here, Squad 5 fireman Bobby Smith uses the spreaders to free a man whose foot was trapped. Cut over the eye, damaged foot or ankle, this guy was using his cell phone with no apparent concern. What isn't known is whether he was on it when he crashed into a light post. This particular tool is the largest of three sizes, weighing about 60 lbs, and applying a force of approximately 18,000 psi.

(Above)

"Ray"

Then Deputy 6th District Chief, Ray Orozco after a fire one cold early Sunday morning, 2/14/03, looking for his long-time buddy Lt. Pat Lynch. Quick thinking, full of stories, tales, and humor. As of this writing, he is the CFD Commissioner, much to the utter delight of the rank and file.

(Opposite Page)

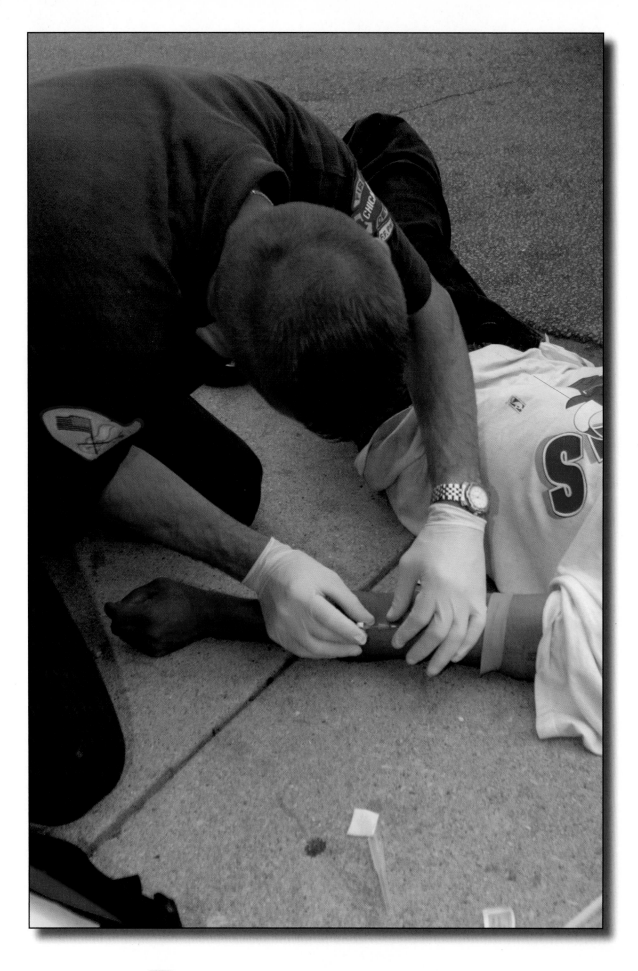

Paramedic/fireman Chris Steinmetz, then of Engine 38,
starting an IV on N. Broadway in June '02.

C ar Accident 7/29/04.
This boy was dragged ¹/₂ block under a car before the driver realized it. In an effort to get the boy out from under, he stopped and backed up, dragging him another few yards. When, he stopped, the boy was removed and some few minutes later help arrived in the form of Tower Ladder 5 and ambulance paramedic Larry Murray.

Fire College: University of Illinois Fire Service Institute in Urbana.

This is one of many classes at a 5-day training for working firefighters, utilizing hands-on experiences; in this class extinguishing propane fires. A large propane tank with a special valve opened for the exercise is used. One task is to concentrate water on the top of the tank allowing a flow of water to run over the sides to keep it cool. This is more effective than pointing the stream directly at the tank. The correct method is to shoot the stream so part of it ricochets off the top. The main task is to slowly approach the point of attack with water, blowing the flames away from advancing firemen until they can get close enough to cap the valve.

Drill at IFSI Fire College.

Ceiling rollover is the extension of the fire plume, or tongues of flame that have become detached ahead of the plume at ceiling level, a recognized warning sign that a fire is rapidly progressing towards 'flashover'. Observe how the flame is spreading across the ceiling. There are hot combustible gases that, when mixed with smoke, may ignite into flashes of flame. When those gases simultaneously ignite, we have a flashover. Despite superb protective gear, a firefighter has less than two seconds to evacuate a room when this occurs.

Tower Ladder 5 at extra alarm at 19th & May, 1/15/04.

(Above)

Pickin' Up.

After the fire is defeated, the crowd of onlookers gone, the media people back in the vehicles and trucks beaming to their editors, there is still a lot of work to be done. Hoses have to be sorted, drained of water and either rolled or put back on the engine hose-beds, ladders have to be collapsed and, if frozen, thawed and placed back on the trucks, hydrants have to be turned off and capped; all this after busting their backs and risking themselves to save people and property.

(Opposite Page)

On 4/11/03, at a 3-11 alarm fire on 22nd St. in the Pilsen neighborhood of Chicago, then Lt. Mark Altman, the youngest of three fireman brothers, fell through a floor into the basement. His men were on the other side of the hole, moving in a smoke-filled storefront. Burned on the head and face, he managed to climb on some file cabinets and pull himself up through the hole. Battalion Chief Jerry McKee saw Altman and asked, "where are your guys?" "I don't know," Altman gasped. The Chief immediately called a "Mayday" and a prolonged blast from a truck horn almost knocked Altman and I over. Mayday is an international emergency code word used as a distress signal in voice procedure communications, signaling "everyone out," a precaution in a dangerous situation and a way to find out if anyone is down or trapped. Altman's and McKee's guys got out through the back of the store. No one else was injured.

(Above & Opposite Page)

5th District Relief Lt. Ed Toomey catching a few gulps of air before going back in at a Still and Box alarm on 41st and S. Prairie, 4/15/05.

"Lucky it wasn't his head," was the remark Capt. Juan Reyes quipped when questioned about the time Truck 48 fireman Angel Reyes' helmet was run over by Engine 38 as it pulled out on a run. Firemen leave their gear on the floor next to their rigs and accidents will happen.
(Right)

To the Roof
(Opposite Page)

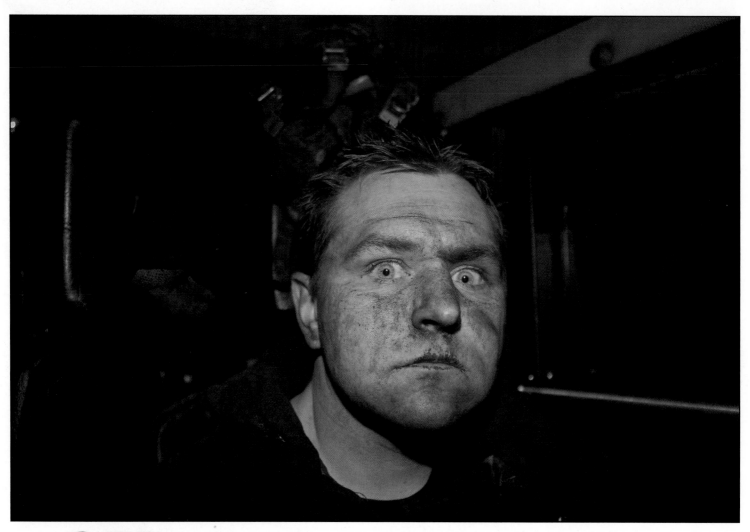

CFD Fireman Marty Halloran, now an engineer, after three fires with Squad 5, back to back.

Capt. Juan Reyes, now a Chief, looking a little nonplussed and wondering just how they were going to lift an enormous woman of close to 400 lbs onto her bed from the floor to which she fell. Out of respect for her privacy she is not shown.

DEPT.

"**Boomer**" Squad 1 fireman Joe Kubik's
kid Anthony. He's always saying "boom," thus his nickname.

SQUAD

1

In the summer of '02 I found my way to "crazy 8's", E-38, T-48, on the West Side, where I learned about the humor necessary to do the job. The man with the orange hat is Joe Tito of Truck 48 clowning after helping with a broken natural gas line. He has one of the largest hat sizes in the CFD.

(Above)

The Great Cheeseburger Eat
The firehouse is a place of great humor and wisecracking, much like a professional sports team. The difference, of course, is that these men (and for some time now women as well) are dealing with life and death on a daily basis. Without humor they would not be able to do their jobs. This was one meal that wasn't interrupted by a fire or medical call. Here, Truck 48 fireman Marty Calkins (now Engineer), former Gordon Tech H.S. Illinois state champion football great and an "all decade" playoff team selection, vacuuming a cheeseburger of his own thoughtful creation on a dare, just as he scooped up a fumble and ran to the end zone in the closing moments of the state championship game, scoring the winning touchdown. He won the eat as well.

(Opposite Page Top)

The Little Peanut Butter Eat
On his last day as a fireman before promotion to Engineer, his wife and children visited Marty at the firehouse.

(Opposite Page Bottom)

A Storz Lock on the rear of Engine 83. These are used to secure hose attachments.

(Above)

This "hand pump" is carried to small fires, e.g. a kitchen stove or a waste basket fire. It weighs about 50 lbs. This, together with an ax, pike pole, helmet, boots, coat, mask, oxygen bottle and harness, weigh well over 100 pounds. Like many objects around a firehouse, it is carefully crafted and beautiful in both form and function.

(Opposite Page)

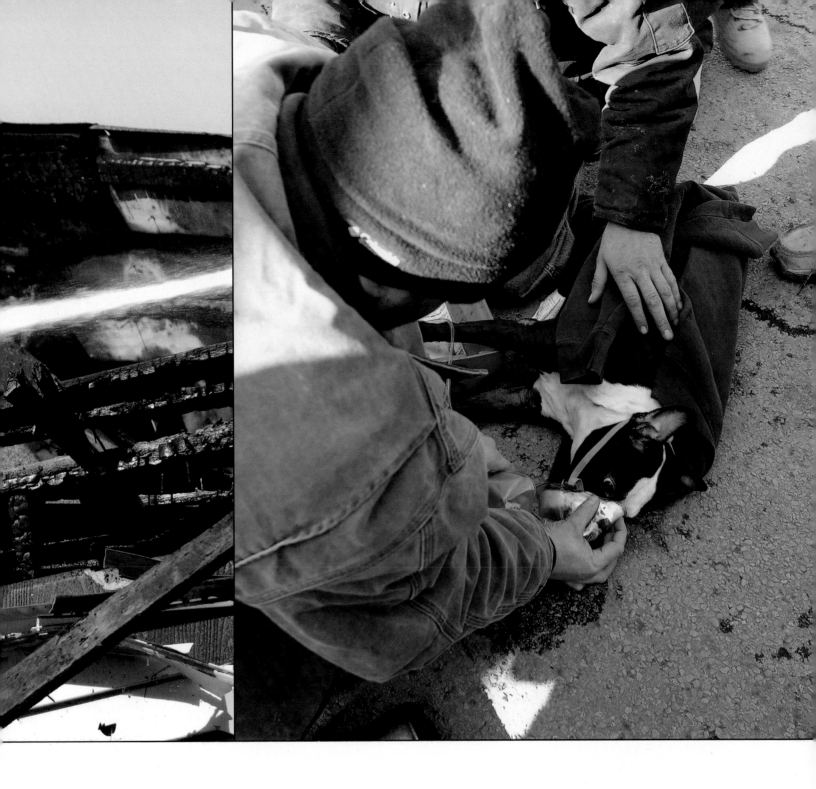

Workin' the Pipe. CFD firemen with their heads through a burned-out roof during a 3-11 alarm fire in the Pilsen neighborhood.

(Opposite Page Top)

Extra Alarm Fire on W. Monroe, 2/5/03. The big guy is Squad 1 fireman Bill Latham (now a 5th District Relief Lt.). To his left is Squad 1 fireman Manny Soto. Other Engine and Squad 1 firemen unidentified.

(Opposite Page Bottom)

After his owner yelled he was still inside, Lt. Jack McKee went back in and pulled this dog out of a fire on 11/30/02. An EMT (emergency medical technician), too busy to administer it himself, provided the oxygen. The man on the left is a friend of the owner on the right. As far as is known, the dog survived.

(Above)

Still and Box 1/30/03 on West End.
Very few of us ever see what these guys go through.
(L-R) Jeff Kovash, Paul Keller and Jim O'Brien.

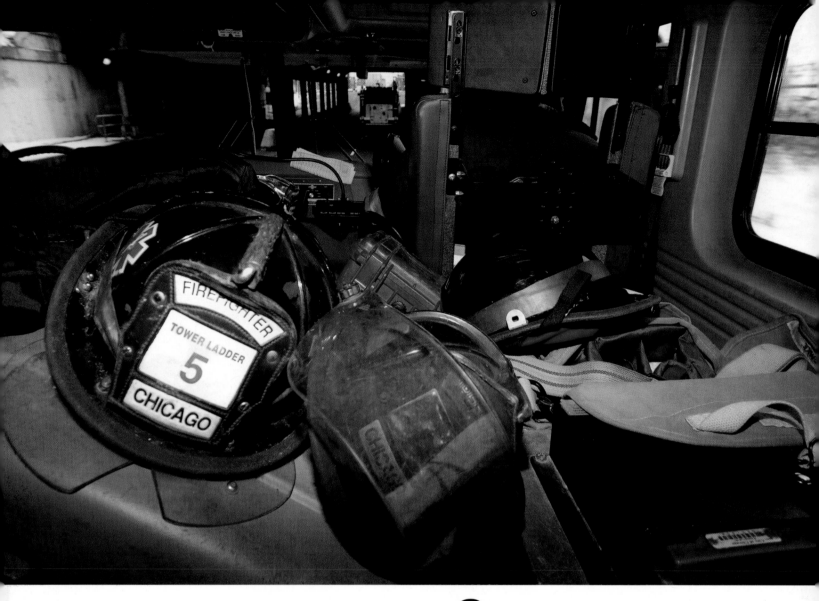

O n a Run in Tower Ladder 5.
Standing on the sidewalk or pulled over to the curb, when these vehicles rumble past, Engine 23 in the lead, and Tower Ladder 5 behind, vibration shaking your teeth, sirens and horns hurting your ears, lights whirring and flashing, engine rumbling and roaring, color blurring in a flash, everything seems excited, very loud, and almost desperate. But inside, it is surprisingly quiet, even calm. The engine always goes first because it has hoses and an internal pump. The firemen must "lead-out" the hose and "make the hydrants" before large amounts of water can be put on the fire. Correct tactical placement is hugely important. Going to, and at, a fire, time seems compressed, but on the way home, on the return to ordinary life, ordinary clock time ˄   ˄. It takes a while for the adrenalin to ˄   ˅ ˅els

(Above)

S mothering the Fire.
CFD Fireman Moe Demus, now a Lt., holding a line distributing a dry chemical agent to an electrical fire in a large power plant. One way to put out a fire is by drowning it with water. However this does not work with class B fires (involving flammable liquids, oils, greases, tars, oil-based paints and flammable gases), or class C fires involving energized electrical equipment. They have to be smothered.

(Opposite Page)

Truck 48 fireman Joe Tito's huge lid.
The encrusted thing a kind of badge of honor, indicating participation in many fires. One time, they saw a guy intentionally rubbing his helmet on a blackened and soot encrusted wall. They waited 'till he was off to scrub it clean as a whistle. Ha, the guy was furious. The message? No Hollywood make-up artists, braggarts, or dishonest showoffs needed or wanted.

(Above)

"Water Carnival" 2/11/03
Firemen use this term to describe a situation where there is so much interior fire they are unable to get into the building for an aggressive attack, something many of them relish.

When a fire is too involved to be fought from within, all kinds of aerial equipment and ground hoses, pouring many thousands of gallons of water, are used. On a very cold night, wind chill (-12°), on W. Washington St., this is the result. When the fire is raging, the building really does appear to be inhabited by a raging beast. Afterward, it sits there like a frozen, dead animal after a winter storm on the prairie, dark, cold… lifeless.

(Opposite Page)

Squad 5 Lt. Cliff Gartner at a still-and-box on S. Parnell.
(Above)

Wash-down.
(Below)

This looks more like a film set than a fire scene at a still-and-box, 12/06/02, on S. St. Louis.
(Opposite Page)

_____

5th District Chief Cortez Holland (L) with Gene Ryan, Deputy Fire Commissioner-Operations, at a 2-11 alarm fire on Chicago's South Side. These men are professionals: calm, clear thinking, with many years' experience – too much happening to pause and observe feelings, but they are revealed in a frozen instant.

(Above)

El Diablo in Pilsen. This was an extra alarm fire in the Mexican neighborhood known as Pilsen, on Chicago's near South Side in January of 2004. Note the horns, and to the left the nose, and below it the mouth and below it the chin and neck; a profile of the beast looking out to the left and laughing, roaring his triumph of destruction. Note the fireman on the ladder at the extreme right.

(Opposite Page)

113

In some neighborhoods in Chicago, it is quite usual for cars to be torched for any number of illicit reasons.

(Above)

Humpin' Horseshoes
Al Duckman, of E-83 at the time of the photo, carrying two "horseshoes": two 50 ft. 2½" lines weighing about 50 lbs each. The nozzle, or "pipe", weighs between 2-5 lbs. They are folded in this manner for fast deployment.

(Opposite Page)

Tower Ladder 5 Fireman Larry Garza's helmet at fire on Washington St., 2/11/03. He had been up in the basket on this very cold night (-12° wind chill).

(Above)

Fireman Nora McDonagh of Truck 3 at an extra alarm fire at the Butterfield 8 restaurant on N. Wells St., January '05.

(Left)

Countless children say, "I wanna be a fireman when I grow up." Here, a friend of CFD fireman John Mogan's young son tries on John's coat and helmet.

(Opposite Page)

F ireman Anne Cacioppo after an overhaul drill on N. Ashland. Anne is also a paramedic and became an RN while working full time on the CFD.

(Above)

4 th District Relief Lt. Pat Curley (L), and Engine 48's Mike Smandra. These guys are intense: serious about their work and highly motivated.

(Right)

CFD Truck 8 Fireman Marty Calkins, dressing a guest in full turnout on Christmas Eve '02. Incidentally the guest is Jesse Jacobs, the author's son.

(Opposite Page)

Opening The Roof.
Firefighters climb on a roof from truck ladders and shimmy along the peak, as if they were riding a bronco, to a spot identified through fireground radio communication and whack away usually using axes, or circular saws until they open it. This procedure is essential as opening the roof allows fire, heat, and gasses to escape. This greatly aids firemen working inside. The first photo with two firemen on the roof shows the beauty revealed even in these harsh conditions. The photo on the right demonstrates just why the roof must be opened. Even if there is no fire, heated gasses travel upward and with heightened temperature, can spontaneously ignite.

(Above & Opposite Page)

Salty-Dog – The Smarty

Salty wandered into Engine 116, Sq. 5's house about five years ago, an emaciated one-year-old and never left. According to some reports, he was starving and close to death. Like many people in the Englewood neighborhood, life was hard. It's no picnic to be a dog on the street in the South Side Ghetto. He prefers to ride in one of the Squad's two trucks. He knows when the speaker calls for Squad 5 and bolts through the swinging door, beating the boys to the rigs; this in a house with an engine, ambulance and, until recently, the Deputy District and District Chief's cars. There are a different number of rings for each vehicle and he knows these as well, not budging when the ones he can't ride are called. One Sq.5 shift doesn't allow him to ride, so he just doesn't respond when they are called. He jumps up to E-116's call. If one Sq. 5 guy doesn't want him to ride while others do, Salty will not go to his door but instead run around to the other side without being told, and occasionally he will just hide in the rig to be there when they pull away. One time, a civilian car broke his hip and the guys took up a collection for the surgery.

S alty... He's there when THEY need him.

CFD
Squad 5: Shun Haynes.

S quad 5 fireman Jerry Rickert.

A n unidentified Chief, and Squad 1 fireman, Kelly Burns, doing what is known as "pulling hose"; here a 2 ¹/₂" line at a building explosion and fire. With water rushing through them, these lines are very heavy and inflexible. How do you get more hose into the building and close to the fire? By pulling it in. You EARN it.

(Opposite Page)

In the "Box"
Back to back with the driver's seat in Squad 5's lead vehicle: O^2 bottle, harness and bag with mask, helmet and gloves. Its hard to describe what these guys endure day in and out. They do it so matter-of-factly, with humor and a kind of detached distance people working close to life's edge develop in order to do their work and stay sane; doctors, nurses, paramedics, police, etc.

Squad 1. The first of two vehicles that go everywhere together. On the floor are Halligan bars, boots, an axe, and a gut belt used to secure equipment 'round the waist. The back of each seat contains a harness with a 45 min. compressed air bottle that snaps into a flexible frame held in the seat. All one need do is pull on the straps and get up, pulling it out of the back of the seat. The bag hanging over the seat with the number 3 contains a SCBA mask (self contained breathing apparatus). The small object on the seat's right side is called a "pass alarm." It displays the amount of air in the tank, and activates a loud, continuous screeching sound if the air runs out or if it is motionless for 18 seconds, this to help rescuers find you if you go down.

Fire ravaged innards.
On the stairs in a burning building, firemen try to stay next to the wall, as they are less likely to give way and collapse sending a fireman down to the floor below or further.

(Above & Opposite Page)

Tommy Garswick, Squad 5.
(Above)

5th District relief Lt. Will Trezek (L) backing up the man on "the pipe" connected to a 2 1/2″ line, at a fire on S. Aida St. 4/10/04. It was first called in as a rubbish fire.

(Background Photo)

Squad 5 fireman Tommy Garswick is getting a boost into the attic from teammate Pat Noonan. Standing by with light is Engine 116 Capt. Sean Burke.
(Opposite Page, Lower Right)

Valuable Equipment at Rest
Actually, Squad 1 fireman John "Scooter" Scheurich squeezed into a shelf, repairing part of the Squad 1 Truck.
(Top)

Angel on His Shoulder
Paramedic Mary Hallock of Ambulance 30 at a pin-in accident, on 85th and Damen, 3/25/05.
(Bottom)

W e humans… so fragile, breakable, vulnerable, and also tough, durable, and resilient… Above all, lucky.

From left, Dan Ryan, T-49; Billy O'Boyle, Sq-5; Lt. Mike Caracottie, E-126 and Ron Kirby, T-49 during overhaul at still and box, 7317 S. Cregier, 4/28/06.
(Top)

These beautiful brass hose nozzles, rarely used now, are still a part of many engines' inventories. On the Left, a "distributer nozzle," the top of which spins as water is coursed through it. The other is a "bell fog nozzle." They speak to the pride and art of craftsmanship.
(RIght)

Epiphany. 1/23/04
(Opposite Page)

Mike's Last Day.

38 years on the job! And then Capt. Mike Cahill, far left, did not want to go. Imagine that in our world. How many of us love our work that much? How many runs? How many fires? How many days? And... how many stories? Funny, sad, technical... you name it, Mike's got a story. One is about the '68 riots on the West Side of Chicago after Dr. Martin Luther King was murdered. The entire sky was aflame, visible to all in this huge city. "We'd come in the morning, drive our cars to the fire, and take over for the guys on the previous shift. They would drive our cars back to the firehouse, and the next day the third of the three shifts would do the same for us. We'd hear over the radio 'go to this block and pick your fire'." It just takes a little something and whammo, a floodgate in his memory opens and out pours the wisdom, funny stories, history, understanding of people, and Irish geography You know Charlie O'Sullivan? Mike has a talent for people; can talk to anyone, on any level: chiefs, firemen, busted-out guys on the street, confused old men and women, Black, White, Latino...makes no difference to Mike. The guys loved him...(from L) Larry Garza, Jamie Gonzalez, Vinicio Espinosa, Rudy Cassanova, Paramedic Kevin Smith, Lt. Bill Block, Paramedic Field Officer Clifford Boyce, Miguel Monzon, Mike Schrader, Dave Bautista (with baseball cap), B.J McGuire, John Fitzpatrick (in shorts), Mike Schultz, Dan Voris, Louie Charne, Jeff Iwema, John Folak, Lt. Paul Viramontes, Dave Reyes, and Roger Engles.

Deputy District Chief Eddie Enright, loved by his boys, on his last day, after 39 years on the job. From right, John Schienpflug, Kelly Burns, Tom Carbonnieau, Joe Kubik, Capt. John Collins, Billy Latham, Tony Budvaitis, Brandon Dyer, Doug Schick, Lt. Bill Duffy, John Haring, Danny Truesdale (barely visible). As one can see, most of them are not exactly happy with Chief Enright's leaving.

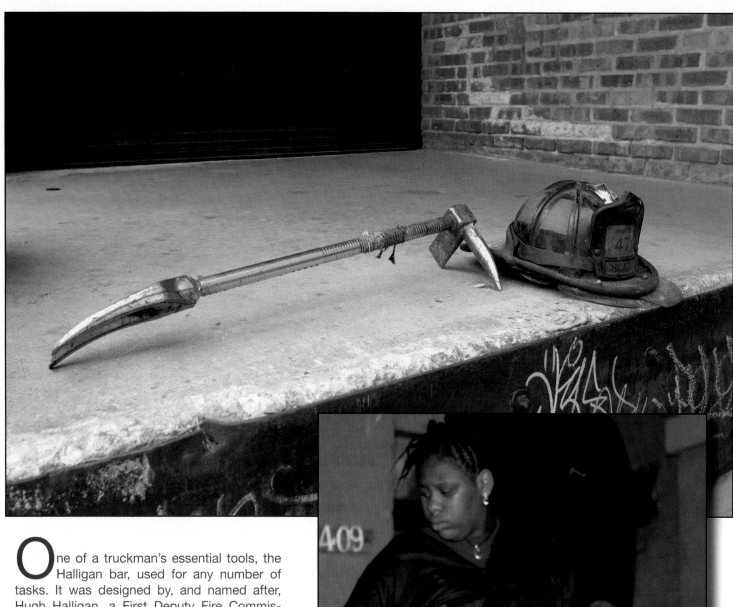

One of a truckman's essential tools, the Halligan bar, used for any number of tasks. It was designed by, and named after, Hugh Halligan, a First Deputy Fire Commissioner in the New York City Fire Department. The Halligan is a multipurpose prying tool consisting of a claw (or fork), a blade (or adze), and a pick, which is especially useful in quickly breaking through many types of locked doors. The adze or fork end of the tool can also be used to break in through an outward swinging door by forcing the tool between the door and doorjamb and prying the two apart. Using a K-tool and the adze or fork end, a lock can easily be pulled. There are many other uses of the Halligan tool, including vehicle rescue and opening walls.

(Top)

Out on the Street
House burning, glass crashing all around from firemen breaking windows in order to cool the hot gasses inside, huge engines and trucks all over the place, lights flashing, engines running, men running all over, connecting hoses, putting ladders to the roof, running inside with them, and these two... lost in it all, cast out by the forces of nature in the middle of winter and fortunate to get out with the clothes on their backs.

(Right)

Pride...
Sq.5 fireman Corey Hojek's axe, polished, sharpened and ready to go. He spent over half an hour grinding, sharpening, and polishing it. Pride in one's tools is a part of the job.

(Opposite Page)

Working Chief
Former 1st District
Chief Tom Donnellon at extra-
alarm fire in January, 2006.

Late stage of 4-11-alarm on 26th and St. Louis, 1/21/06. The Streets and Sanitation Department was called to free the sewers.

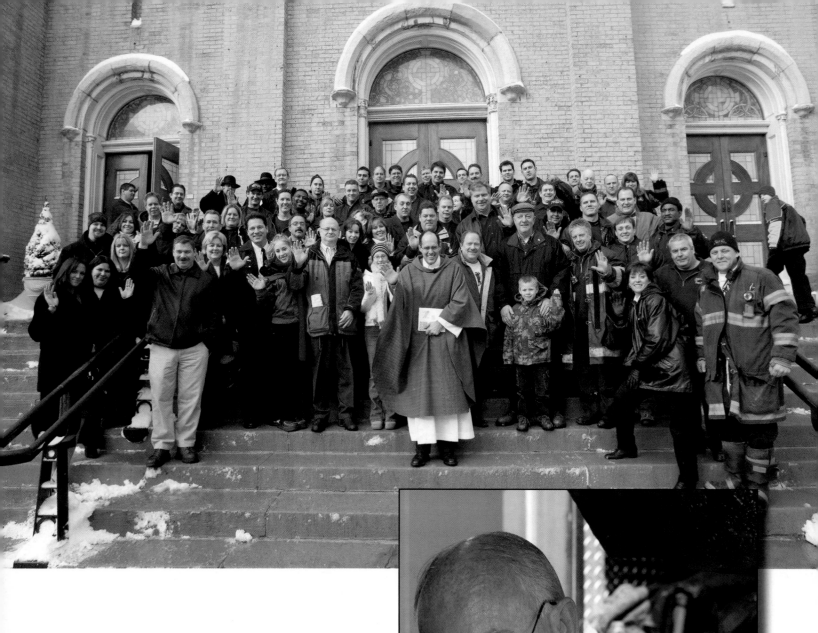

One of the guys on Squad 1 was in the hospital diagnosed with leukemia. He was not afraid to say he was frightened. The photo at the top was taken after a mass for him. That really boosted his spirits. Arty got well and married his log-time girlfriend, Denise (front right, black leather jacket). Firemen stick together, help build each other's homes, do the electrical, the plumbing, fix each others cars, on and on… In many ways they are an extended family.

(Top and Right)

Overhaul, one of the last stages of defeating a fire, when they continue to poke, pull, and chop walls, ceilings, floors, doors and window frames searching for hidden flames or smoldering material that will eventually ignite, requiring a return to the scene.

(Opposite Page)

Art Noonan

Making the Roof
A cold night, -5°. The limits of endurance are multiplied when the job simply has to get done. One cannot say, "oh well, I'm too tired, I can't do it."

(Above)

Part of a drill at Illinois Fire Services Institute.

(Opposite Page, Top)

Fire Drill at University of Illinois Fire Service Institute in Urbana. Open steel racks are stuffed with wooden pallets and stoked to actual heat and smoke conditions. This gives firemen some experience with what it is like in a real fire.

(Opposite Page, Bottom)

Shun Haynes

"Jammer"
Squad 5 fireman Jamar Sullivan, during overhaul, 2/17/05.

Squad 5 fireman Brian McArdle during overhaul at still-alarm, 7/9/04.

SS1 1979

Snorkel Squad 1 in 1979, a precursor to today's squads. (L-R) Tom Murphy, Bob Hoff, Jerry McKee, Bill O'Boyle, Steve Bybee, Tom Donnellon, and Tony Thomas. These men, and the others of that era, set a high standard for the squads of today. Tom Murphy became a Lieutenant, Bob Hoff, an author and as of this writing Assistant Deputy Commissioner–Operations, Jerry McKee, now Battalion 14 Chief, Bill O'Boyle, now retired, was Captain of Squad 1, Steve Bybee is Chief of Support Services, Tom Donnellan was an Assistant Deputy Fire Commissioner before his retirement in 2005. As of this writing he is 1st Deputy Commissioner of The City of Chicago Department of Buildings. *Photo courtesy of Robert Hoff.*

(Below)

Little Bob Hoff at five years old with his Dad, Captain Thomas A. Hoff at an inspection in Soldier Field in 1960. Captain Hoff was killed when a building collapsed over him during a fire at 70th and Dorchester in Chicago, Feb 14, 1962. The story of Capt. Thomas Hoff, Bob, and his older brother and fireman, Ray, served as the basis for the film "Backdraft." Chief Hoff has been concerned with firefighter safety and survival his entire career co-authoring a book "Firefighter Safety and Survival" with Lt. Rick Kolomay, another CFD fireman's son. Looking up at this photo on his office wall, CFD First District Chief Bob Hoff said, "I look at that picture every day, and it continues to drive me, to serve as my motivation to be the best I can be…. Every day."
Photos courtesy of Robert Hoff.

Captain Thomas Hoff

Battalion Chief Robert Hoff

Rescue 11/28/86

Here, the same little boy from the previous page, carring a 68 yr. old woman from a fire building on N. Sheridan Road. The fire was in the basement and Bob Hoff found her in dense smoke on the 1st floor, somewhere in the middle of the apartment. Many thanks to long time and knowledgeable fire buff and photographer Jim Regen for permission to publish his photo here. Bob Hoff is a two-time winner of the Labert Tree Award for valor.

Photo © Jim Regen, 1986, 2006.

a pile of smoldering material. The roof had collapsed the night before. It was learned later that this fire was started by two young men who had been arrested for shoplifting, were out on bail and, as an act of vengeance against the store owner, torched the place.

151

The Dive Team

This is a separate unit but it is also something that the squads do, water rescue, in full scuba gear. Here, Brian Hurn (above) and Mike Vasko (left) during a drill in the freezing waters of Lake Michigan.

T he photographer at 5°

154

rashing, tearing, ripping, banging, lots of voices and other racket is what it's like inside during overhaul. Here, Squad 5 fireman Pat Noonan is using a pike pole to pull parts of the wall and ceiling down looking for the source of smoke.

The Sun Times fire, 12/29/04.
Former 1st District Chief Bobby Hoff (L), and former CFD Commissioner Cortez Trotter. Hoff is now Assistant Deputy Commissioner – Operations.

(Above)

Tony Budvaitis, Squad 1.
When it is five 5° below everything become more difficult: walking, swinging an ax, putting water on the fire, pulling hose. It takes superior levels of determination and endurance to do whatever is necessary to get the job done. Yet, they don't want to look at a lot of it. "You there, just do it." This work often entails some degree of physical suffering, even injury, to say nothing of the emotional toll. Still, most are ready with a wise-crack or a joke, and the next call.

(Opposiite Page)

Through the Sidewalk
A fork lift with its trapped operator; his leg wedged between the sidewalk and the lift's retaining bar. Looking at the camera, is Squad 5 fireman Glen Keyes, who eventually got the man out by cutting the retaining bar. The man with the dark sweater and mustache is his boss, Special Ops Chief Mike Fox, and to his left, is Deputy District Chief, Sylvester Knox. The grey and yellow supporting device is a telescoping Paratech air shore strut, stabilized by shooting compressed air into it. All this for one man.

(Above & Opposite Page)

Engine 116 firemen Cristina Sarnowski (work-ing "the pipe"), and Jimmy McDermott at still-and-box on Chicago's South Side 11/27/05.

(Above)

Washdown

– CFD Fireman Kevin Helmold, Engine 54

When civilians see this photo, many ask: "Is that what it's like inside?" I explain this is after the fire is basically out and the procedure is to make cer-tain it stays that way. Possibly embers could be glowing or even a small flame, usually in a remote spot. And I answer, "yea that's what its like... after it's over." This is quite different than Holly-wood films, which do serve a useful purpose, but cannot depict actual conditions. When the fire is raging, the conditions inside are black. Often they are able to get down and crawl at floor level, and see across a narrow space, about 1 ft. high.

(Opposite Page)

This is a "pin-in extraction" of a young woman trapped in her car, the roof of which has been removed using a variety of hydraulic powered tools known as The Jaws of Life™. This woman's femur was broken. A white backboard is being positioned before the victim is removed from the vehicle, a standard procedure before placing a victim in an ambulance.

(Above & Right)

Ready to Roll
Squad 5 fireman Tommy Garswick's stuff.

(Opposite Page)

Level 1 Hazmat, at building explosion near S. Archer Ave., 3/10/06. There are three levels of Hazmat: 1 investigatory for possible trouble; 2 exposed hazardous substances usually requiring Hazmat protective suits in order to go in; 3 very severe. The pile of rubble and bricks smoldered for quite a while, eventually requiring the use of foam. Fireman on "the pipe" is Chuck Miller, Truck 48.

(Above & Opposite Page)

4-11 on Cermak Rd., 1/21/06. These "large diameter" 4″ hoses are used between a hydrant and an engine, or from one engine to another, to a tower ladder, or snorkel.

(Above & Opposite Page)

A propane drill at University Fire Service Institute, in Urbana, IL, during a five day course in firefighting for working firemen offered as "Fire College." This time they learn to cautiously approach a propane tanker, pushing the flames in the opposite direction. Once close enough, they cap it.

This is another part of the class on extinguishing propane fires. Here, propane is shot out of a pipe on a stand and the firemen slowly approach with a fog nozzle.

Fire Monster in Repose
After the beast has been extinguished, the main is retracted and lowered, and water drained from the spray nozzle on the underside. The nozzle used for the flames is just to the left of the "5" on the front.

Lt. Jack McKee at 56, having recently returned to the Chicago Fire Department after a 20-year absence. As a member of Snorkel Squad 1 (SS1) Jack was badly injured when a brick wall collapsed on him. Not fully healed after a year, he was dropped by the CFD. He went to work as an electrician for NASA at Cape Canaveral helping with, among other things, setting up emergency landing sites for the Space Shuttle as far away as Africa. After 20 years at the Cape, the World Trade Center and the Pentagon were hit. He decided to see if he could get back on the CFD. In his mid-50's, he passed the physical test and got back to the traditional family job with his old rank of Lieutenant on Tower Ladder 5. When asked why he decided to return, Jack McKee said, "Because I felt like I owed something to the job."

(Above)

Tower Ladder 5 at 4-11 alarm fire on 26th & St. Louis in the Pilsen neighborhood, 1/21/06.

(Opposite Page)

Working Chief
City Wide Relief Chief Pat Knightly pulling some hose into a room with similar conditions to those in photo to the right.

(Above)

Not at all like the Movies...
In order to show firemen fighting a fire inside a structure, movies such as "Backdraft," or "Ladder 49" must film them with fire, but without one of the most lethal elements, smoke. In reality, rooms are filled with thick, toxic black smoke containing such lethal chemicals as hydrogen cyanide, carbon monoxide, hydrocarbons, and various others toxins depending on what is burning. Sometimes all but a foot or two off the floor are enveloped. Imagine this photo without the light - not much to see. Initially, firemen have to crawl on the floor to have any visibility at all, heated smoke rising as it does. I read this to one of the guys and he quipped, "If you think this scene is bad, remember this is after the fire was extinguished."

(Opposite Page)

oe Kubik, Squad 1.

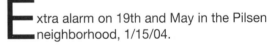
xtra alarm on 19th and May in the Pilsen
neighborhood, 1/15/04.

ngine 122 fireman John Janiga pulling hose at still and box on S. Parnell
12/18/05.

(Opposite Page)

Eyeing the Beast.
Sean Devine of Squad 5 at 4-11 alarm fire on 4/11/03.

(Above)

Engine 116, Lt. Bob Griffin (L) and fireman Todd Taylor surveying the scene at a pallet warehouse and yard fire on Chicago's South Side, 4/10/04.

(Opposite Page)

Lenny Pitts

Engine 116 fireman Cristina Sarnowski inside for overhaul 11/27/05.   *(Above)*

Squad 5 after a fire on S. Garfield Blvd. 12/8/03. (L to R) Marty Halloran, Lt. Will Trezek, Corey Hojek, Steve Groszek, Joe Atkins and Tommy Meziere.   *(Below)*

Jo Jo and Bobby
Firemen Joe Atkins and Bobby Smith (in rear) of CFD's Squad 5 after a still alarm on the South Side.

"Wash-down" is part of the last phase of defeating a fire. "The pipe" is the heart of firefighting. No water… no victory. It is part of the initial entry, used to push flames and hot gasses back from advancing firemen, and then to push them out ventilated spaces, windows, holes chopped in the roof etc. It is usually inside from start to finish. *(Background Photo)*

W ho's that guy next to Calkins?
Unit citation photo after a ceremony in the Chicago City Council, 1/16/03. (front row, L-R) Lt. Mike Smandra, Angel Reyes, Moe Demus, Chicago Mayor Richard M. Daley, Marty Calkins, Tommy Lunz, Jimmy Piscalo, and Capt. Juan Reyes. (Back row, R-L) John Veller, Denijal Milat, Scott Shawaluk, unidentified. They received the citation for pulling two young children and an older man out of a fire. The men who did this were Lt. Mike Smandra of Engine 38, and Firefighters Moe Demus and Angel Reyes of Truck 48.

(Above)

I nside Engine 38 with Engineer Larry Ulansky.

(Opposite Page)

3/11 alarm on Chicago's South Side, 7/12/04.

(Above)

Tower Ladder 37 clearing some branches for fire attack at a large apartment building, 7/12/04.

(Opposite Page)

183

Extra-alarm fire on S. Wentworth Ave. in Chinatown, July 4th, 2003. The buildings are part of a long line of row houses under construction. Three were involved, two seriously. A disgruntled homeless man whose refrigerator-box home had been taken away by the contractor started the fire. Sometime later he was apprehended.

(Above)

I showed this photo to a District Chief from a Boston suburb. He said, "Jeez, I have been looking at these things for over 20 years and never realized they were this beautiful." CFD Engine 83.

(Opposite Page)

(Above)

Sometimes while riding with these guys one wonders just how a problem will be solved. In this case, a car turned on its side in the middle of a boulevard. With no hesitation they just pushed it back to an upright position. They possess many different tools for different jobs, but in the end, with or without them, it is intelligent and cautiously aggressive action that gets the job done. When complimented they usually laugh and shrug and wonder where I have been all my life.

Engine 83 fireman Steve Andolino Sometime in the Fall of each year, every hydrant is opened to ensure they are properly flushed for the coming winter. Failure to do so could result in a frozen hydrant. Other parts are also checked for workability.

(Opposite Page)

The Fortuna Rapid Deployment Craft (RDC) CFD Squad 1 fireman in ice drill using the Fortuna or "banana boat" in Chicago's Monroe Harbor. It provides excellent buoyancy and rescue capabilities for ice, cold water, surf, flat-water, and swift-water rescue. The RDC's floor is open at each end, facilitating rapid and easy entry. It is stored in a 2' cube, inflates in under 1 minute using a SCBA air bottle, and can be carried by one person.

(Above)

Like father like son. Billy O'Boyle of Squad 5 during overhaul of a 2nd floor front room on Chicago's South Side. His dad, another Bill O'Boyle was one of the men on SS1, the forerunner of today's Squads (see pages 152 & 153).

(Opposite Page)

CFD Squad 5 fireman Glen Keyes, looking for hidden remnants of the beast. 2/17/05.

The Chicago Fire Department's Memorial Park. Here, the 544 firemen who have died in the line of duty are honored and remembered. An actual helmet and boots were bronzed.

Squad 1 Lt. Bill Duffy helping lower a 35 foot ladder at a still and box on 21st & Albany in the Pilsen neighborhood, 12/6/02.

The Ice Man Cometh...
(This at same fire as opposite page)

(Above)

4-11 alarm, 2121 W. Washington, 2/11/03. 9°, 20-30 mph wind for many hours. The fire building had to be attacked from outside using many different aerials, that is, towers and snorkels with men in enclosed platforms (baskets), and truck ladders. The result was a lot of spray flying all over the place.

(Opposite Page)

Two men working in an apartment courtyard failed to properly shore the sides of the hole in which they were working and the sides collapsed, trapping one of the workers in a thick, viscous mud to the top of his thighs. A trench rescue procedure was instituted by 14th Battalion Chief Jerry McKee (not shown) with firemen from Truck 48. This specialized form of rescue involves shoring-up the sides of a trench, and digging a trapped worker out. Very often, they don't use many high tech tools though, occasionally, sophisticated air-shores and other devices are employed. Many of the tools are simple; wedges made of wood, sheets of plywood and shovels. Power tools are generally not used, as these may cause injury to a trapped victim. Plywood sheets are placed along the sides of the trench, and 4x4 wood bracers are placed between the sheets. Wedges are driven in behind the plywood to force them together against the bracers. This, to prevent further collapse of the ditch sides. Once the ditch is secured, the rescue team may enter the ditch and recover the trapped victim.

(Above)

Jo Atkins

"Pickin' Up"

After the fire, hundreds of yards of hose have to be drained of excess water and put back in the engine hose bed. Some of it is also rolled.

(Opposite Page)

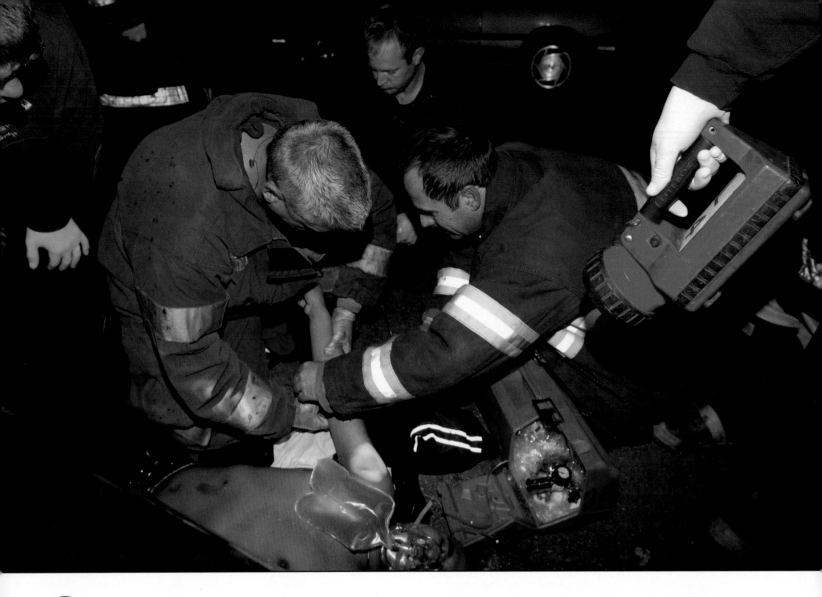

Gang warfare, often drug related, is not an uncommon eruption in the Pilsen neighborhood despite the vast majority of honest, hard-working residents. This young man was shot in the upper right quadrant of the abdomen in the area of the liver.

(Above)

After 9/11, the outpouring of caring and respect for fireman increased many times over, especially with school age children. Perhaps their teachers felt this as well. They brought these kids to Engine 109, Truck 32, Ambulance 34 on 9/19/02.

(Opposite Page Top)

The engines have the hose and nozzles. After they have controlled the fire to some degree, these guys step in. These firemen arrive on trucks and await the order to go in. Standing with their various tools, axes, Halligan bars, and pike poles, they remind of medieval pike-men waiting to do battle. The man holding the day-glow red light is an officer. Over his shoulder is a thermal imaging camera.

(Opposite Page Bottom)

Vicki Hernandez

We put our hands together to thank you for your hard work.

A very cold day, around 9° with below zero wind chill. Engine outlets and controls continue to work, even in very cold conditions.

(Above)

F ireman/paramedic Rob Nelson of E-83, after a fire on a hot night in '02.

(Opposite Page)

CFD Squad 5 fireman Tom Garswick after a fire where four young children died.

(Above)

Garswick stripping window frame during overhaul phase of still alarm on Chicago's South Side 2/17/05. He is held in high regard by many in the CFD.

(Opposite Page)

Jerry McKee

Index